HAVE WE FOUND OUR BETTER SELVES?

(WHAT WE CAN LEARN FROM COVID-19)

MERIL R. SMITH

Contributing Writers

Linda Ullah

Rosanne Johnson

Maria Thompson

Anthony Cedoline

Lynn Chen

Israel Golden

Leah Smith

First Edition

ISBN: 978-1-63945-089-3 (Paperback)
        978-1-63945-090-9 (E-book)

The views expressed in this book are solely those of the author and do not necessarily reflect the views of the publisher, and the publisher hereby disclaims any responsibility for them.

Writers' Branding
1800-608-6550
www.writersbranding.com
orders@writersbranding.com

I have been awed and inspired
by the thousands of first responders,
health care professionals, medical support personnel,
teachers, and essential workers of all kinds,
who have put the health and safety of others
before their own.

They are true heroes in this time of COVID-19.
I am also awed by the sacrifices
of the families of our heroes.
In many cases, families have not seen their mom
or dad, wife or husband for weeks at a time.

Finally, I am awed by the millions and millions
of families who have had to suddenly
shelter in place and whose life was
literally changed over night.

May this book serve as a guide for families
who have survived the COVID-19 pandemic and gained
insight into their better selves. The book is filled with
stories about people, young and old, which have taken
time to examine what is really important and stepped up
to help their fellow citizens and their communities.

Hopefully, this book will help you reflect and seize the
opportunity to also find your own better self.

Life lessons learned during the COVID-19 pandemic:
Be safe, wash hands, wear masks and practice social
distancing! If you do not like wearing a mask, you are really
going to hate a ventilator.

TABLE OF CONTENTS

Surviving A Pandemic

Families Growing Together

Compassion, Empathy And Helping Others

Younger Age Children

Daddy Led Activities

Small Businesses

Try Thinking Inside The Box

SURVIVING A PANDEMIC

The "New Normal" Be Damned!

For some people, this may be a shocking statement: "As much as you might like, we are not returning to the 'old normal' of 2019."

The vaccines are helping American turn the corner on COVID-19, if we are lured into a false sense of security that COVID-19 is on the way out, we are damn fools. COVID-19, in one variant or another, will be around for the foreseeable future. If we are not vaccinated, we are an "infection" waiting to happen.

> *"It is okay if you are anti COVID-19 vaccine shot.*
> *It is okay if you do not believe in science.*
> *Remember, after thousands of years,*
> *"Survival of the fittest" is still working well."*

The COVID-19 pandemic has created a reality that has resulted in many changes in the way most of us live. Some of the changes we have experienced may no longer be necessary, while others will become a part of our everyday "new normal."

- Surviving shelter in place, more than once
- Washing hands again and again
- Staying six-feet apart from others, especially grandparents
- Isolating from family and friends
- Shopping for groceries
- "Working from home" or not being able to work at all
- Using Facetime with grandparents, family, and friends
- Learning to use ZOOM to be connected with groups

- Knowing someone who has died from COVID-19
- Watching the death toll from COVID-19 tragically soar to numbers, which has exceeded the total number of U.S. deaths in the last five wars.

There are other new normals that are not quite as apparent, but will be around for a long time. Some are easy to see. Some may make us laugh. Others take some thought.

- How we view toilet paper. Having toilet paper has certainly taken on a new importance for a lot of people, including me.
- The large number of Amazon Prime delivery vans we see every day. My wife recently commented that she sees more Prime Vans on the streets than Tesla's.
- Grocery and restaurant delivery services are not going away. They are here to stay.
- The impact of ZOOM and how we communicate with groups of people.
- The way local, state, and federal governments need to respond to emergencies. No elected official, at any level, imagined the crushing impact COVID-19 would have. Other kinds of mass emergencies will also happen: floods, droughts, hurricanes, tornadoes, and new epidemics. Perhaps, as individuals and governments, we will actually learn to think and plan ahead. The question about large disasters is not "if," the question is "when?"

The COVID-19 pandemic has taken a huge emotional, social, and economic toll in the United States. Many of the ideas on which we have based our daily lives and daily activities were suddenly, and dramatically altered, almost overnight.

If you do not believe in the possibility of outside events creating "new normals," let me give you just one very simple and concrete example that has continued to change our lives for over twenty years.

- When was the last time you went to the airport to catch a flight and you did not have to take off your shoes?
- When was the last time your luggage at the airport was not scanned or searched?
- When was the last time you did not have to leave for the airport an hour early, just to be sure you would be able to get through security and not miss your flight?

The terrorist attack on September 11, 2001 resulted in the immediate grounding of every aircraft in the sky. Literally, overnight a "new normal" was created. Flying has never been the same and will never go back to anything resembling the "old pre September 11th normal" in our life time. Like it or not, everytime we fly, we are directly impacted by a "new normal," something for which we did not ask nor want. If you are under twenty years old, taking off your shoes is your "normal." For the rest of us, taking off our shoes is a part of a "new normal." The creation of a "new normal" for flying is something for which we did not ask or want. We still have to take off our shoes.

Let me be blunt: How old were you when you had chicken pox, mumps, or measles? *Really?* If you do not believe in a "new normal," or if you choose to resist or deny every change created as a result of COVID-19, **STOP READING IMMEDIATELY AND DO NOT BUY THIS BOOK!**

If you are bothered by or concerned by changes in our daily lives, you need to read about ways you and your family can adapt to a "new normal" and actually come out better. There is nothing medical in this book about facing a "new normal." People do not have to like the idea of having to adapt to a "new normal." What counts is how each of us adapts.

Let's step back for a moment. Within three months of COVID-19 becoming a reality, a new book, **"Riding The Waves During A Pandemic,"** was written. Few people took the book seriously and we experienced new, more deadly, waves of COVID-19. Despite vaccines, the COVID-19 is still here and waves are continuing around the world. Like it or not, COVID-19 will be with us, in some deadly form or variant, for years to come. What we need are lots of ideas and ways to help families and youngsters learn to better survive emotionally and socially in times of great emergencies. As individuals and families, we can actually come out better human beings because of what we learn from COVID-19. Coming out better is what this book is all about finding our better selves, now and for a lifetime.

Although COVID-19 has changed the way we do many things in our lives, it does not mean we are helpless. If we choose to learn and adapt to future new normals, this terrible pandemic may actually become a blessing in disguise for most of us. If we choose to learn from our pandemic experiences and put what we have learned into action, our family and our children can emerge with a stronger sense of self and family as well become better citizens. In the process we may just learn that each of us has more power than we ever imagined to make things better for ourselves and those around us. We might even find ourselves looking at family and community with a whole new way. We may find ourselves more caring towards family members and develop stronger bonds within our family and supporting our community than we could have ever imagined.

INTRODUCTION

The COVID-19 changed our lives almost overnight. We went from life as usual to sheltering in place, washing hands, keeping distance from others, and wearing facemasks. We went from hyperpartisan politics to phoning neighbors to be sure they were alright. We went from adults going to work and kids going to school to almost everyone staying at home all day, everyday. We went from our normal concerns about health to washing our hands many times each day. Although many of us have lived through mass tragedies — the World Trade Center, wars in Afghanistan and Iraq, as well as hurricanes, tornados, floods, and wildfires. None of us have ever experienced anything like the COVID-19 pandemic. The United States has 5% of the world's population and 30% of the world's COVID-19 cases.

As a child, I remember hearing stories about the 1918 Pandemic. So many people were dying in Philadelphia that every morning, bodies were placed on the sidewalk and picked up by priests in horse-drawn carts. The bodies were taken away and often buried in mass graves. In Philadelphia, 759 people died in a single day. Later, I remember reading that one third of the world's population became infected.

In 1918, there were no such things as vaccines or antibiotics. However, there were common sense public health measures. What was done in 1918 may sound a bit familiar.

- People were told to isolate themselves if they had any symptoms.
- People were told to liberally use disinfectants on surfaces to kill germs.
- People were told to practice good personal hygiene including washing hands.
- The size of public gatherings was severely limited.

- Many stores and shops were closed to prevent the spread of the virus.
- In places where there were a large number of sick people, areas were quarantined.

The measures in 1918 worked well and the rate of infection slowed down. Soon, people thought the worst was over and restrictions were relaxed. No one expected a second wave to hit. The second wave of the 1918 Pandemic was far worse. Over 675,000 people in the United States died and between 20 and 50 million people died worldwide.

Many people thought COVID-19 would end quickly and life could go "back to normal." The COVID-19 vaccines are certainly making a huge difference in the number of people becoming infected and dying. COVID-19 is well into its second year and will be infecting people, although at a slower rate, for a long time. Going "back to normal" is just not going to happen. Some of the changes we have experienced as a result of the COVID-19 pandemic are becoming part of our everyday lives—like it or not.

The big question for many people is *how are we going to adapt?*

There are people who will deny the realities of COVID-19, "sit in the corner, suck their thumbs, and have an ongoing rant." Other people will look at the realities and simply accept the ongoing changes resulting from COVID-19. A third group will examine how COVID-19 has changed their lives and the lives of their children and learn, change how they think and perhaps become better human beings and better role models.

As Americans, we have faced many difficult situations, unexpected events, and tragedies. We have a history of making learning from them and making adjustments. There are also people who will examine any situation and look for the opportunities that come from them. Looking for opportunities in the era of COVID-19 is what this book is all about.

Years ago, people were captivated by the ideas in Gene Roddenberry's *Star Trek*, including many teenagers and young adults. The Star Trek stories were more than science-fiction. They were filled creatures from other planets, with stories about history, with storylines filled with imagination, challenges and possibilities. The characters on Star Trek frequently used handheld communicators to talk with each other. Although they looked real, communicators were actually props made of balsa wood and painted silver. These simple balsa wood props captivated the imagination of some young fans and were the inspiration to create and develop something most of us carry with us every day, the cell phone. What are possibilities that may come out of the COVID-19 pandemic? Possibilities that may better our lives and the lives of people around us?

LIFE IS NOT ABOUT
HOW MANY TIMES WE FALL.

LIFE IS ABOUT HOW WE GET UP!

Yes, the COVID-19 pandemic is a tragedy. Millions of people have contracted the virus and hundreds of thousands of people have died. Certainly, there were big mistakes and failures at many levels during the first year of the pandemic. We also know the big challenge is how we get back up from the pandemic and move into the future.

There is no better example of learning how to stumble and get back up than the thousands of seniors who graduated from high school. The pandemic turned the normal high school experience upside down; gone were in-class learning, going from class to class each day, social events, dances, sports, hanging out with friends, proms, having a support system of friends, teachers, and counselors with whom they interacted with every day.

When it would have been easy to quit, teens have dug down and found the inner strength to overcome the kinds of obstacles no other students have faced. With distance learning, teens had to take responsibility for logging in for each class, participating in discussions, learning the material, completing assignments, and getting assignment sent in to be graded. In the pandemic, teens have learned to be more responsible, more resilient, more independent, and more determined. Teens had to learn ways to take control of their education and their future. Almost all seniors rose to the challenge. What they have learned about themselves will help them deal with challenges they face during the rest of their lives.

The following eloquently reflects the new normal for high school seniors. Unfortunately, I do not know the source.

Every generation has a defining moment,
a collective experience. This is ours.
It is not what we lost. It is what we found.
When we get knocked down, we get up!
Impossible is nothing!
We will push forward!
We will fight like hell to create the future we want.
Our class survived the Covid-19 pandemic
and is tied together forever.
Nothing will ever break our bond.

We all stumble and fall. We make mistakes and bad things happen. Sometimes we play major part in bad things that happen in our lives. Other times, we play no part and events are thrust upon us. How we get up when we are knocked down and how we move forward with our lives is far more important than stumbling and falling.

- Are we bitter?

- Do we learn nothing and make the same kind of mistakes again and again?
- Do we consistently blame others for our woes rather than take our share of responsibility?
- Do we take what we have experienced and learn from it?
- Do we use the experience to become better human beings for ourselves, our families, and, perhaps, our community?

High school seniors rose to the challenges they faced. What they have learned about themselves will help them deal with the challenges they will face during the rest of their lives. These teens have learned that life is not about how many times we fall. Life is about how we get up. All of us can certainly learn lessons from our teenagers.

There is nothing medical in this book and that is intentional. This book is focused on ways that can help individuals adjust to a "new normal." Each section is broken in two parts. The first part of each section is a short overview of ideas and things to ponder. The second are true stories about people, how they coped emotionally, how they took positive actions, and how they grew and became better human beings.

FAMILIES GROWING TOGETHER

Wash Hands, Wear Masks and Practice Social Distancing!

Months of on again off again "Shelter in Place" can be enough to make any family a bit "crazy." Some families have continuous upsets and hurt feelings. Other families seem to have found ways to grow together and make their family bonds tighter. These families seem to focus on finding positive ways that involve each other. Often family members play or work together, learn new ideas, create new experiences, find common interests, support each other on projects, and share chores.

This kind of thoughtful and caring approach by parents is not an accident, it is intentional. Parents model what they want to see in their children. Parents set the stage for making bonds between family members a lot stronger, especially between siblings.

Now is the time to think. What is really important to us? How can we change the way we live in ways that will help us all become better human beings?

An Old Memory: A New Reality
(Linda Ullah)

An Old Memory

In the early 1990's my learners at Edenvale Elementary School in San Jose, California embarked on a local history project. Their goal was to write a book about the history of South San Jose. In their research, youngsters found about local Native Americans and the mysterious healing pools located in nearby hills. Research also included the first European families that settled the Santa Clara Valley and what happened as the land became farms and ranches.

As a part of their research, youngsters found out that the first school in the area was built about 1850. Oak Grove School had a small number of students from first through twelfth grade. The school was next door to a stagecoach stop, where teams of horses were switched while weary travelers quenched their thirst with cold beer. At lunch, older students would sneak over to the stagecoach stop and would return to school slightly tipsy. This lunchtime routine caused quite a stir with the wives of the local ranchers and the next year Oak Grove School was moved a half mile away.

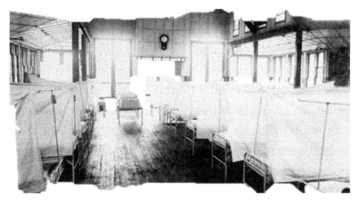

(Source: Oak Grove School District archives)

Pictures were also found of the 1881 and the 1904 Oak Grove Schools. There was an unexpected photograph of the inside of a school. What was this picture about? When was it taken? Was the school being used as a hospital? More research was required. Youngsters reviewed all sorts of information, including years of attendance registers. The big surprise came from 1918 attendance register. All the students were marked absent. A note in the attendance register said that school had been closed because of the flu. More and more information was collected and then became part of a chapter in The Great Edenvale Detective Caper. The fifth and sixth grade students involved in writing the book had never heard of the 1918 Pandemic. Even the principal, who grew up locally, had no knowledge of this part of local history.

It is now 2020, about thirty years later. My mind flashed back to the early 1990's and the excitement we felt as we researched, discovered amazing stories, and wrote The Great Edenvale Detective Caper. I never thought about the 1918 pandemic as anything more than long ago history and a good story. We are now in the middle of the Covid-19 Pandemic and suddenly I found myself devouring a book by John M. Barry, The Great Influenza, a book about the 1918 flu! Who knew?

A New Reality

On March 13, 2020, Jay Inslee, the Governor of Washington State, announced that schools would be closed. Everyone was directed to shelter in place. The local Spokane YMCA was immediately closed, and youth sports were canceled. The local news that night included a story about the first three cases of Covid-19 in Spokane. Spokane is in the eastern part of Washington, far away from the big Seattle metropolitan area. With local cases, the COVID-19 pandemic became real for all of us living far, far away from Seattle. The next day we began to self-quarantine. What are we going to do?

We started by going on walks and hikes just to get out of the house. Technology has helped self-quarantine easier. My sons have reconnected, via Zoom, with cousins living all over the world. Our YMCA started real time exercise classes. We also found Tai Chi videos for more home exercise. I learned to use Instacart for online grocery order and delivery.

We Google Duo video chat with our son, Isaac, his wife, and our two-year-old granddaughter, Talia. For my grandson, Ethan, who lives in San Diego, we had a ZOOM Birthday Party. Isaac teaches at San Diego State University. Leah, his wife, teaches at an alternative high school. Both are finding it very challenging to teach virtually with an active toddler scurrying around the house.

Noah, our second son, a realtor, is working virtually from home. Stephanie, his wife, runs a homeless "warming shelter" in downtown Spokane. Her job is high risk for COVID-19. She has not been able to come home since sheltering in place started. The two oldest boys, thirteen-year-old Ethan and eleven-year-old Logan, were rebellious about having to stay home. Seven-year-old Cohen has enjoyed playing outside, when schools were closed for the remainder of the year and virtual classes began to really hit the two older boys hard.

To curb Ethan's and Logan's anger and rebelliousness, we gave them research projects. They researched COVID-19. In addition, they did projects about two past pandemics/plagues of their choice. The projects helped them better understand why we are self-quarantining. Thankfully, their attitudes are improving. Ethan and Logan are now running two miles a day, taking hikes, and helping with household chores. It's really hard for them not to see their mother.

My son, Noah, does not really know how to cook, so both grandmothers take turns cooking dinners and leaving them on the front porch every evening. Seven-year-old Cohen broke my heart when he called to say he misses my hugs.

When my new glasses arrived, the optometrist's secretary brought them out and placed them on the hood of my car. When I lost a crown from a tooth and had to go to the dentist, I carried wipes and wiped down all the surfaces in the dentist's office. Twice, I have gone out to purchase plants for my garden wearing masks, rubber gloves, and taking wipes. In Spokane, we are fortunate that the few stores that are open have early morning hours reserved for senior citizens.

Our family continues to make adjustments and find new ways of surviving together. We have put our passion for traveling on hold. I am glad the Eastern Washington weather is improving and I can still garden— one of my pleasures in life. We have adjusted to a new normal. I do not know how long the pandemic will last or what life will be like. My hope is that perhaps we human beings will have become kinder and more tolerant.

Life Lesson: *No matter the results of the COVID-19 pandemic, we will be living in a new normal. Will we go back to a highly polarized and angry country? Will we learn from the experience and realize we are all in this together? We have the opportunity to be kinder, more tolerant, and more caring to our fellow Americans.*

Sheltering = Bonding
(Lynn Chen)

"Sheltering in Place" is something we are actually going through! Through my years of schooling and work, we have prepared and practiced for all sorts of emergencies. Never did we think about how to care for ourselves and others in a worldwide pandemic. Many people had never even heard the word "pandemic." "Epidemic,"—Yes. "Pandemic,"—No! The COVID-19 pandemic hit us hard, very hard. "Pandemic" has become a term known around the entire world. For sure, the COVID-19 pandemic is something that will be written in history books for our grandchildren and great grandchildren to learn about!

Through the COVID-19 pandemic, some positive things did happen. As humans, we usually only think of ourselves or those that are close to us. What I have learned through this ordeal is about the love and care that everyone around us has shown. It's actually true caring. We have neighbors, friends, teachers, educators, co-workers, and bosses checking in on us making sure that everyone is safe at home and taken care of. How amazing it is to see the true side of people with whom we usually don't socialize. We have created bonds by socializing more than before, however in very different and new ways.

As a family, we have documented many sides to how we function now. It's amazing to see when we go grocery— shopping and distance ourselves at least six-feet from one another. It's amazing that people are only allowed to go into the grocery store a few at a time. Many shelves are empty. Toilet paper, paper towels, or household products are all sold out. Masks are required and gloves are highly recommended when we have to go into a store.

What we miss most is seeing our families, my parents and siblings. It's been months since we visited relatives. What is also amazing today is we have technology to help support our daily communication with them, even from far away. There is no dining in restaurants, only take-out. One thing that has amazed me is to see how well almost everyone has integrated so many new things into their daily lives.

Kids have switched from going to school to online learning; what a new way to adapt education! What saddens me most is not seeing my eighth-grader be acknowledged for his educational accomplishments, or see him walk across the graduation stage. I feel for all the seniors and graduates who will not have a proper sendoff, senior ball, honor night, graduation ceremony, and celebrating with families and friends near and far.

It's hard knowing that our annual trips with friends will be cancelled. Our annual summer trip will also be cancelled. Mother's Day was the first time we have not celebrated with our mom's and grandmother's. It's hard when birthdays are only celebrated among the six of us. Something about sheltering 24/7 is just plain hard!

What has been the most amazing part of the weeks of sheltering in place is seeing my four children bond. Their age gap has always put a distance between them. However, seeing that they are playing together in the backyard, really talking to each other and, yes, sometimes beating each other up are goals we have reached. They have bonded and really care about each other.

Finally, a picture I would like to share is the four of them getting ready to plant a garden, pulling out all the weeds, going to the stores to pick out the seeds, fertilizing the soil, and finally planting their own vegetable seeds.

All four of them are awaiting the seeds to sprout. It's not about the plants—it's about the bond they are creating among each other— just like the seed hoping that one day it will sprout, grow, have flowers, produce vegetables and see them harvested and shared. What a special way to really end our family experience with this pandemic.

> **Life Lesson:** It is possible for four kids to bond together in ways I never imagined. And that is a very good outcome!

Shelter in Place Experience
(Rosanne McGrath Johnson)

Even as I type this, it still seems like this is all just a bad dream or like "I'm stuck in a movie." Our families' world literally came to a halt weeks ago and we are still adjusting to our new normal!

Never have I wanted to homeschool my children... NEVER! My son, Hunter, has always wanted me to homeschool him because he says he "hates school." However, after the first few days of homeschooling, Hunter kept saying, "I just want to go back to school!" Now he is finally getting his wish and he wants to go back to school. Crazy, I didn't get it! When I asked him why, he responded, "I thought being homeschooled meant you get to sleep in, only work for an hour, and then get to play and eat snacks all day!" Needless to say, this new home centered school environment was not what Hunter was anticipating.

Thankfully, all my kids have amazing schools and incredible teachers. Still I put in several hours helping them, making charts to keep them on task and organized as well as submitting assignments online. Charts are a huge help in trying to keep three kids, in three different grades, and three different schools stay organized. I have a daughter who is a junior in high school, a son who is in middle school, and a son who is in third grade. Fortunately, I also have a mom who is a teacher and who can be called upon if additional help is needed. I am happy to say that in third and seventh grade common core mathematics, "I'm killing it!" I now regret all the times I said, "I am never going to need to know how to do this math formula in real life."

We all thrive on a schedule. The first week of shelter in place the kids didn't really have homework so I let them relax and adjust to our new life and schedule (or lack thereof). After the first week, I made an individual schedule for each kid. The schedule included when they had to be awake and up, chores, a list of every school subject and the school work that needed to be done for that day. Time was also scheduled for exercise and free time. Creating a schedule for each kid was a blessing and has made things run smoothly. I don't have to say a thing. The kids wake up, check their schedule, and get to work. Not having to nag anyone has been wonderful—it's the new normal. I add some fun things each week, such as family game nights, cooking together, family walks, and a family "Chopped" cooking challenge!

My husband and I are considered "essential employees" so we both still go to work. I am thankful that my kids are old enough to stay at home alone, otherwise that would be an additional hardship. My husband is in the habit of calling me on his way home and asking, "What are our plans tonight?" We normally have a busy schedule with three kids involved in four different sports, so we are constantly on the go. We miss the baseball games, volleyball, flag football, and the trap tournaments. Now when my husband asks, "What are our plans tonight?" Now I say, "NOTHING!"

(Source: Roseanne Johnson)

It felt weird as I went through our family calendar and began deleting everything. The most painful ones were deleting our plans for a huge family cruise, a trip of a lifetime, with my siblings and their families as well as plans for our anniversary... everything cancelled! We celebrated our 18th wedding anniversary, quarantine style. A card table and folding chairs were set up. I made a delicious dinner for two, complete with a table cloth and special table settings, and a romantic dinner on our front lawn. This anniversary was definitely one for the books!

At the beginning of April, school was cancelled for the rest of the year and it looks like so will everything else. When I watch the news and see the devastation that COVID-19 is causing, I am grateful that we are not sick. I am grateful that my husband and I still have jobs. We are considered essential workers. I am grateful that our family is together, sheltering in one place, and that we are looking out for each other.

I am OK with 2020 when the world stopped and everything was cancelled. Years from now, I will be able to say to my family, "Remember how our lives were changed and what we learned about ourselves and each other. Remember when we had to stay home all day and not play with our friends so we would not get sick or spread the deadly COVID-19. Yeah, those were some CRAZY times!"

> **Life Lesson:** *We never know what we can really do until we are put into a very different situation. Coping sometimes takes imagination and creativity. I have definitely learned that everyone in my family and my large extended family can rise to the occasion.*

"Izzy"
(Israel Golden)

My name is Israel Golden, but I have never gone by that name. Everyone I know calls me "Izzy." I am considered an "essential worker." Normally, I work as a cashier at the supermarket, but right now, during these weird times, I am also the "doorman." I keep track of how many customers that go out at the exit door or how many can be let in. I am also charged with performing a continuing headcount to make sure we are staying under our designated capacity. Why me? I worked as a bouncer in bars for 14 years, and I got chosen for my experience of working with difficult people. Most customers are very nice and understanding. Every once in a while, someone is a real jerk and makes demands to be let into the store anyway. My job is to try to keep them calm while not letting them in. Another part is making sure each customer is wearing a mask. It's hard to understand why someone would be so crass as to put people or themselves in danger of contracting COVID-19.

With help from the Edenvale Elementary School's Adopt-A-College Scholarship Program, I am also attending San Jose City College. It took a lot of years for me to finally realize that I want and need a college education. We are using video conferencing called "Zoom" through the canvas-online app. Since the class is interactive, I must say it is closer to being in class than an ordinary online course.

My English 92 class is wonderful. I am able to communicate with my professor, Professor REW as he likes to be called, and participate in class like I normally would. With some of the scholarship money, I purchased a nice computer video and microphone system. I really appreciate to be able to take interactive classes and see my classmates as well as the professor.

My commute to work and back is mostly by bicycle since the bus and light rail systems are now limited and do not connect well to where I work. My life is pretty much normal because I am still working. When I am not working or commuting, I am at home doing school work, writing for my English class mainly and preparing for summer school classes.

On Tuesdays, I really miss going out and studying while I treated myself to a meal at Sweet Tomatoes. Recently, I experienced my very first college Spring Break. I had great plans that suddenly changed because of COVID-19. I really missed going to the beach. I really wanted to spend Spring Break on the ocean, surfing the waves, and relaxing. Instead, I am making the most of it, by doing my English homework and writing stories. I am also rereading comic books. I am planning a way of using my new camera and microphone to read and post comic books online so that elementary age kids can do something unique during sheltering in place.

I have found a couple of "positives" that have come out of the shelter in place order. First, I have learned more patience in helping people. Everyone is upset and stressed. People can and will take their bad day out on you if you let them. I have learned to not let customers get me down. Instead, I have taken the proactive choice in being happy and helpful. It is my decision, and my decision alone, whether or not to be happy or unhappy. I have learned to keep other people's negativity from drowning my positivity. Secondly, I have learned to do my schoolwork online. When I started college last year, I told my counselor that I did not want to take any online courses. When I was in school last, in 1995, there was no internet in schools. I told my counselor that was the way I learn. Well, I guess I was wrong. As a result of COVID-19, my English-92 class was switched to an online conference/lecture class. I am currently doing all my courses online and that I have been able to adapt and overcome my fear of online learning.

> **Life Lesson:** *Sometimes we are surprised at the ways we can change when life-altering events happen. I am pleased when I have learned this about myself. I can adapt. I can change. It's my attitude that makes the difference.*

FROM A TEACHER
(Maria Thompson)

As I received word that our school would be closed for a while, my immediate reaction was "YAY!" My first graders had been "squirmy" and had "Spring Fever"— no pun intended. In my mind, a week off of school couldn't have come at a better time in the year. A couple of days before, I had sent an email home to parents. I said that I had noticed the students were having difficulty staying focused and completing work in a timely manner. However, I also said, this is the time of year we usually see this type of behavior and it is par for the course.

With a sigh of relief, I knew that the parents would have to deal with their child's spring time behavior for a few days. I truly thought of this as a welcomed break. We will be back in school soon and everything will be back to normal.

As the first week went by, I began to realize that this was not going to be as easy as I expected. The students at my school in Walnut Creek, CA were sent home with a journal and 5 library books. The first week the students were instructed to read daily and write about what they read. In addition, they were told to work on math using an online platform. Students were to complete one math lesson a day. That was it!

As we found out more, I realized that school was going to be closed, at the very least, for three weeks. The second week we were briefed about the many online platforms we would be using, none with which I had any experience. For each new platform, there was daily training that came with a huge learning curve. We would be working through our own learning curve, while teaching our students at the same time.

Suddenly, we now needed to know how to use and navigate Google Meet, Google Classroom, Seesaw, Raz-Kids, Screencastify, Literacy Footprints, Google Chrome, Google Chrome Extensions, Freckle, Epic, Audible, Dreambox, Bridges Math, Google Slides, Google Forms, Google Docs, and Google Sheets. I think I forgot a couple of names.

Teachers not only had to prepare lessons on these platforms, but we also became the computer teacher (IT Director) and troubleshoot students' computer issues. Honestly, I would not have spent the time learning all these platforms without being required to shelter in place. We were instructed that we would also have three Zoom Meetings per week.

My immediate reaction was how do I keep the attention of 24 squirrelly first-graders virtually? Then I thought, how do I know that each student is using a device that has access to Zoom? In addition, will they know how to navigate any of the platforms? All of these thoughts swirled in my head as the next problem loomed. How was I going to teach important concepts virtually? Moreover, how will I know my students understand new concepts?

As a first-grade teacher, and a year where students are required to learn to read, I wondered how I will be sure that my struggling students will have enough intervention at home to reach the grade-level benchmark. Will the parents know enough to help teach reading? Will there be any way that I can, somehow, teach reading through the computer?

I spent many nights up until 2:00 AM thinking of how this online learning would work and how I could reach my students during this tumultuous time. As a result, I continued to spend approximately 14 hours a day creating and teaching online lessons that I hope will be successful. In addition, teachers also have staff meetings, district grade-level and school grade-level collaboration meetings each week.

Our school grade-level team spends approximately eight hours a week on collaboration, working out ideas and techniques we use for our remote teaching. I continue to fine-tune my lesson plans, to make them easily assessable to students and parents alike. On Zoom, I meet daily with students in small reading groups. Also, there are one-on-one Zoom meetings with my four struggling readers. While focusing on the lesson, I try to make Zoom meetings fun. I will tell funny kid jokes, or sing a song, or do a GoNoodle dance while teaching. Encouraging students to be active participants in lessons and sharing what they are learning are important parts of being a first grader.

We do wiggle breaks as needed, go through our days' lesson plans on Seesaw and, at the end, we have a fun scavenger-hunt as their exit ticket. I try to make online learning fun and entertaining while focusing on the skills that need to be learned.

There is another side of teaching with Zoom. There are times where students unmute themselves and you can hear all the noise in their home. I have seen kids jumping on couches, beds, doing TikTok-dances while listening to the lesson, and walking through the house. I have seen and heard siblings argue and some background distractions I'd rather not share.

As a result of teaching online, I have created and implemented an incentive program. I provide positive reinforcement with virtual stickers and virtual rewards. The students earn a point for each assignment turned in, regardless of mistakes, and earn a sticker for every five completed assignments.

A Zoom lunch is offered when a student earns 200 points. Seven of my students have already earned lunch. We eat our lunches together on Zoom and play games like "I Spy" and "20 Questions." We have all loved having this special one-on-one time together! I honestly was a bit shocked that this idea was so well-received. The rewards of seeing my students' smiling faces and of them sharing their homes, toys, and rooms have been wonderful adventures that I would not otherwise have had the opportunity to experience.

What about other areas? Science experiments for my students to try at home are posted on YouTube. Virtual field trips, weekly read-aloud video recordings, daily math lesson videos are also created. We have also had special guests at our meetings such as our Science Specialist, Music Specialist, and Librarian.

I am planning for a virtual talent show for the last week of school. Students will perform their talent virtually for their classmates during our Zoom meeting. With their parent's permission, I will make a personal delivery to each student's home on the last day of school and deliver a gift of masks and gloves. Of course, six-feet social distance requirements will be observed.

Through this pandemic, I have learned that I am genuinely appreciated. My students and their parents gave me an incredible parade of cars, parents and kids, signs and decorations galore, celebrating their first-grade teacher. Families have sent me gift cards, delivered flowers, wrote sweet messages in chalk in my front yard, sent emails, texts, photographs, pictures, letters, edible bouquets, chocolates, balloons, and most importantly they have imparted their love and support as we travel this journey together.

Teaching is hard, but the benefits far outweigh the challenges. Seeing the light in a child's eyes is worth more than all the money in the world. During this time, my students have expressed to me that they are scared and want to get back to school and normalcy as they once knew it. So, do I. I have missed seeing the amazing growth that takes place the last couple of months of school. I will miss celebrating Open House and showcasing student work.

I realized how much I love my job and the importance of interacting with students in the classroom. I know how crucial our role is as their teacher and how I have the ability to impact lives forever. The classroom is a "sanctuary" for learning academically and socially. A school classroom is something that simply cannot be replicated in a virtual classroom.

There is no job I'd rather have than being a teacher of elementary school-age children. I'm pretty sure COVID-19 is an experience that will be etched in these students' minds forever, something never to be forgotten.

I am truly fortunate. I recently was told that I get another year with this same group of students. I am able to "loop" with them and be their second-grade teacher. As we learn together, make up some lost ground, we have the opportunity to continue to touch each other's hearts and lives. As a result of the COVID-19 pandemic, there is a bond that has been created that will link us forever.

BROTHERS BONDING
(Leah Smith)

Ari is seven and Brahm is two. COVID-19 really changed our daily lives. Staying home was something new. Distance learning was something new. Having a seven-year-old and two-year-old together almost all the time was something very new. Day-to-day with the boys has been anything but easy. I am exhausted.

How do I turn this difficult situation into something positive? After thinking about what was really important, I decided that creating a positive bond between the boys would help them build a strong lifetime relationship. Brothers being lifelong friends is very important to me. What might happen if I listed the interest and skills of each boy and see where there might be an overlap? After all, five years is a big difference in age.

Ari likes	Brahm likes
• Sports	• Eating
• Throwing and catching	• Dancing
• Climbing	• Facetime with grandparents
• Being a part of a team	• Books...love books
• Books	• Looking at pictures
• Reading	• Being read to
• Mathematics	• Climbing
• Art	• Going for walks
• Music and dancing	• Waving to people
• Nature	• Blowing kisses
• Going to the park	• Giving hugs
• Being friends with lots of kids	• Learning new words

I know that each child needs his/her own space and time, as well as special parent time. Setting a few minutes aside several times a day for the two boys to be together seemed like a workable idea.

At first, I spent a lot of time modeling how to be an older sibling. Ari soon learned that how he treated his younger brother would come back to him. If he was bossy, little brother would be a bossy toddler. If Ari took his time and was excited to show Brahm something new, it would be met with cheerfulness. Their shared love of books was a great beginning. Brahm loved to pick out books and ask Ari to read to him. Ari would read. Brahm would turn the pages. When something was funny, they both liked to laugh together. Reading together also is helping with Brahm's speaking vocabulary.

(Source: Leah Smith)

When we were able to go to the park, Ari showed Brahm how to climb up playground structures and slide down together. When music is playing, the boys like dancing at the same time. Playing with toys or rolling a ball are also shared activities. I try to pay attention and not over do it with their together time. Staying at home because of COVID-19 was the genesis of the "Brothers Bonding." Supporting the growth in their relationship is something that will we will continue long after COVID-19 is over.

COMPASSION, EMPATHY AND HELPING OTHERS

Wash Hands, Wear Masks and Practice Social Distancing!

There is little doubt that there will be new variants and additional spikes, especially in the area with low vaccination rates. How fast individuals, communities, and officials reach will be the key to how bad each spike ends up. Most of us know what to do: wash hands, wear masks, and do social distance. The question is whether or not we are smart enough to get ahead of new outbreaks.

What did people do when we first started gathering together again? Did we continue new hygiene habits? Did our economy reopen safely? Were federal and state government officials too cautious or overly exuberant? Would there be measures in place to mitigate subsequent waves? Did citizens simply fall back into old habits? Or did families take advantage of what they have learned while sheltering in place? In short, has COVID-19 permanently altered the way many of us think and behave or will we quickly return to the "old" normal? These are all important questions and we have not done well. As COVID-19 explodes around the world, will we learn and do better?

Living through the COVID-19 pandemic has revealed many unexpected results. Many moms and dads are at home. Youngsters no longer attend school. Members of families are spending far more time together than they ever imagined. Information related to the Pandemic is on the television just about twenty-four hours a day. Sheltering at home has totally changed how many families interact and function.

Hearing stories about how individual families have adapted is fascinating. Across the nation, non-medical people have come up with creative and innovative ideas that help in our fight against COVID-19. In doing so, adults and children have the opportunity to learn about, experience, and develop our better selves. The stories in this section share instances of compassion, empathy, and helping others. Perhaps the stories will spark ideas of ways we can help each other get through new waves of COVID-19 and be better for it.

> *Life Lesson: The COVID-19 pandemic has changed our lives. Later in life, many stories people will tell will either be framed as before or after COVID-19. There is no doubt that people, including youngsters, will remember the COVID-19 pandemic for the rest of their lives. As parents, we have this time to show our children how we care about other people (people we may not even know), how we show compassion, and how we show others appreciation. To make these lifelong values, we need to have discussions within our families. Even more important, we need to model these values day in and day out for our children. That's the real way we create lifelong values and attitudes.*

ADULTS MAKING A DIFFERENCE

Wash Hands, Wear Masks and Practice Social Distancing!

When a crisis occurs in our lives, it is easy to react emotionally and negatively. Stepping back and looking at our situation with some perspective is far more difficult to do. This section about adults making a difference is to share a variety of ideas of what can be done in a crisis that can help us and our families survive and survive well.

For most adults, COVID-19 was a direct hit. Within a few weeks, as the pandemic grew, how adults went about their daily lives was "chopped, diced, and pureed." Some people lost their jobs. Others worked from home. Others homeschooled their kids. A big challenge for many parents was to learn how to navigate "distance learning." Most adults had to develop new ways of functioning. Many people starting to reassess their lives, make big changes and ask ourselves what were the things that were really important in life? Obviously, the answer varied with circumstances. However, the question was still asked.

For people with young children, distance learning, homework, and being at home twenty-four hour a day created immediate adjustments. Just setting up work (computer) stations for each child was a huge challenge. Learning how to help each child learn to use distance learning was another huge challenge, especially with different children in different schools.

In some families, both parents are able to work from home. With some creative planning, while one parent is online working, the other parent is supervising the children during distance learning.

Families have found that the limitations of distance learning have created many more conversations and problem solving between youngsters and parents. There is no question that being in a classroom is far better than being isolated and doing distance learning. However, I have been amazed at how families have developed new ways of coping with children not going off to school.

A large number of adults started working from home. At first, working at home seemed a bit strange. Spending an hour commuting to and from work has been a bonus. A year later, working from home has become an accepted practice with both workers and businesses. The practice has been so successful, that some large businesses are actually reducing office space and cutting costs.

COVID-19 has had a tremendous effect on senior citizens. Not only were senior citizens at a much higher risk of contracting the virus. An active social life is a must for seniors and much of their social life came to a screeching halt. For many seniors, social interaction is the most important part of daily life in staying physically and mentally healthy.

On one hand, seniors have had to be extremely careful so they would not catch the COVID-19. For many seniors, they have felt very isolated and disconnected. Other seniors have stepped back and used their life and work experience to come up with new ways to stay connected and be involved.

The following stories are just a few examples of what creative seniors have done to work around the limitations imposed during the pandemic.

LIBRARY BOOKS FOR SENIORS

In many places where seniors live, residents have been required to shelter in place for extended periods of time. Even as sheltering in place is lifted, many seniors will remain reluctant to venture out into places where there are groups of people.

Many seniors like to read and frequently check out books from their local library. Not being able to go to the library has been a challenge. One library thought about using their bookmobile, sitting in the parking lot, to provide books for seniors. The bookmobile idea ended up as not a viable solution to the problem, but it did generate a variety of ideas.

How about bringing books directly to seniors? Librarians and a small group of volunteers came up with a solution. They developed a simple survey and mailed it to seniors who used to regularly visit the library as well as senior living in senior or assisted living facilities. The survey collected information about what genre of books each person prefered, favorite authors, and areas of interest. Every other week a half dozen books are hand picked for the reader and are delivered. At the same time, books from the previous delivery are picked up. The system really works!

Obviously, the project is more involved simply picking up and delivering books. Using each person survey, volunteers pick out books, put them in book bags, put a newsletter or order from inside, and arrange the book bags for delivery. Everyone is required to wear a mask. However, as restrictions are eased, it provides the seniors a much needed social contact with the volunteer who deliver their books. This is a winning idea that can be copied by libraries in many towns and cities.

COME IN FROM THE FREEZING COLD

Unfortunately, the COVID-19 pandemic has not been kind to many people who are homeless or living on the edge of being homeless. In many communities, churches or other organizations open their doors on freezing nights and provide shelter, cots and bathroom facilities for twenty to twenty-five people.

Usually, supervision is set up on three to four hour shifts starting at 7 P.M. through 7 A.M. A supervision team is composed to two people per four-hour shift, traditionally retired people. Providing support and supervision usually requires eight to ten volunteers per night. The need for shelter on freezing night did not diminish, but the availability of seniors to serve on supervision teams fell dramatically.

Fortunately, working age adults heard about the need and responded. Some volunteers were people who no longer had jobs. Other new volunteers had jobs where they worked at home and could adjust work hours. Senior citizens took on the responsibility of providing crockpot dinners and breakfasts, but without younger adult volunteers, many people would have been forced to sleep outside in below freezing weather.

Without the COVID-19 pandemic, many new volunteers would have never considered helping people in need on subfreezing nights. As a result of COVID-19, more people are thinking beyond their own family and responding to broader needs within their community.

KNITTING FOR THOSE IN NEED

For homeless people, winter presents very significant and dangerous challenges. Is there a place to stay when the temperature goes below freezing? When it is not below freezing, how can people keep sleeping outside to keep warm?

A very talented group of senior citizens knits and crochets wool hats, hand warmers and scarves for the homeless. Much of the wool yarn gets donated by community members. When combined with donated wool socks, people sleeping on the streets have a much better chance of keeping their extremities warm and safe on cold nights.

Before COVID-19, this caring group of women met socially once a week to knit and crochet. Throughout the pandemic the group continued to knit and crochet separately because the items they made were still needed. As restrictions are eased and it becomes safer to meet in small groups, these volunteers are hopeful to restart meeting together once more.

Having a sense of humor, this special church group of seniors volunteers call themselves the "Knitwits." As one member told me with a grin, "Knitwits" seemed a far better choice for a name than a proposed alternative, "Happy Hookers."

MOVIE GROUP GOES VIRTUAL

With children grown and out of the house, a group of friends chose to go see a movie every month. Sometimes, the friends went as a group and other times as individuals or a couple of couples would go. At the end of the month, everyone would have seen the movie. Then the group would gather, have coffee, and have a discussion about move. What was the point of view in the movie? What they liked and did not like? How well was the movie acted? What did each person take away from the movie? The activity was like a giant movie review among good friends.

With the onset of COVID-19, going out to the movies stopped. However, it did not stop the group. Each month, a movie was still chosen, but now they viewed the movie in their own homes using "on demand." At the end of the month, instead of meeting in someone's home for a discussion, the group learned to use Zoom.

The importance of this story is that the "movie group" found a way to keep enjoying movies and maintain important social contacts and friendships at the same time.

"NEWS TODAY" GROUP ZOOMS

For years, a group of twenty-five to thirty adults interested in politics and current events met together once a month. Each month, a different leader chooses two or three topics or news stores to present and discuss. As one can imagine, the discussions were always lively, especially ones having to do with politics. With COVID-19 restrictions, the group quickly switched to ZOOM.

COUPLES MAKE ADJUSTMENTS

Whether a couple has been married for forty years or five years, the changes in how people live as a result of COVID-19 put stresses and strains on marriages. Most people have work lives and social lives beyond their marriages that take up significant portions of each day and some of the stress were not apparent prior to COVID-19. The following story is one example.

Often when people are dating, they "can't get enough of each other." After a few years, even in the best marriages, the feeling is not exactly the same. Sheltering in place, being with each other twenty-four hours a day, has put a strain on a lot of relationships. Working out differences as they arise, rather than letting them fester, is more important than most couples realize.

A long-time friend is super organized and approaches almost everything with lists and dates. Her husband has a very different approach. He is much more laid back and lists are not a part of his lifestyle. However, he still gets what he needs to be done.

Suddenly having to be together twenty-four hours a day brought the differences in style out in the open. Clashes in style occurred frequently and there was tention between them; she wanted him to be more task and list oriented, and he was very content with getting things done in his more laid back way. Both had had professional lives where their individual styles were not a problem. However, stress created by weeks of sheltering in place was something new and unexpected in their relationship.

What to do? Obviously, their great relationship and long marriage was not something they wanted put at risk. However, something had to change. In the end, my friend simply said (easier said than done) that trying to get her husband to think her way was not going to work and doing things his way would make her crazy. Instead of thinking that her way was better than his, she chose to simply work at accepting the differences in style. She realized that her style was not better than his. The styles were just different and worked for each person. Making such an adjustment in thinking was not easy, but certainly reduced growing strain and was well worth the effort.

"MORMOR" AND THE BOYS

When you are a single working mother with two young boys in school and COVID-19 hits, what do you do? You cannot go to work. The boys cannot go to school. At first, everything you normally do stops and you shelter in place with the boys.

The school district scrambles, buys and distributes thousands of Chromebook computers, teachers take crash courses in how to use zoom and other new programs, and, after a couple of weeks, actually start distance learning with students. The learning curve for distance learning is steep for both teachers and families. Sometimes, the boys were not able to connect to their classes. Other times, there were problems in completing assignments. Her older son, on occasion, tried to skip distance learning and play online games with friends. (Oops, not a good idea!) For this family, the transition was a bit easier because the mother was very computer savvy and was able to help her sons learn how to navigate this new way to attend school.

When some restrictions were finally lifted, going back to work was welcomed. The workplace was a bit strange; co-workers were wearing protective gear in addition to latex gloves and masks. Instead of the waiting room, clients waited in their cars and were called on their cell phone when it is their turn.

Being able to go back to work, bring in some money, and pay bills felt great. However, a six-year-old Izzy and eleven-year-old Liam together at home alone were simply asking for all kinds of trouble. Fortunately, the number of COVID-19 cases in this small town was very few and the number of people in her "contact bubble" was very small. Mormor (grandmother) came to the rescue.

Asking 78-year-old Mormor, who does not have a computer and much less knowing anything about Chromebook to supervise two active boys, was a huge request. Mormor loves her grandchildren and knew her daughter needs to go back to work.

The task of working with two boys was going to be very challenging for this 78-year-old. Setting up ground rules and writing them on a poster was a must.

- Until the rules change, everyone wears a mask
- Distance learning and Chromebook for each boy are in separate rooms
- When there are breaks between classes, exercise is a must— taking a walk around the block, running the obstacle course, and jumping on the trampoline

- Getting out of the house for lunch and going to one of several restaurants with drive thru windows
- Phone numbers for students to call if they are having a Chromebook problem with distance learning
- Writing mom a note if there is a problem or if assignments are not complete. "Hell has no fury like a mom who comes home and finds one of her boys has not completed the day's school assignments."
- A list of activities and games that they all can play when distance learning is over for the day.

After a year and a half, her grey hair is greyer. After "pulling out her hair" on a regular basis, her hair is much shorter. Mormor deserves an award. Perhaps, instead of a "Golden Globe," she should receive a "Silver Globe" award.

LITTLE FREE LIBRARIES

The idea of Little Free Libraries has been around for a dozen years. Until recently, I had never seen one. However, since the advent of the COVID-19 pandemic and many public libraries being closed, the number of Little Free Libraries have exploded.

The idea of a Little Free Library book sharing boxes in a community is simple. Gather a number, 20 or 30 adult and children's books of different genre, create a container that is attractive and water proof, and install it in a place that is easy for people to reach. People "take a book" and "leave a book." When a Little Free Library gets established in a community, the amount of use and number of book rotated can be astounding.

(source: Meril Smith)

When looking at photographs online of Little Free Libraries, the ideas, creativity in design, and placement is amazing. This is a project that can be done by an individual, small group, or family. There is a lot of information on the web. Little Free Libraries are found in front of homes, churches, schools and other places that are easily accessible.

Little Free Libraries is such a neat way to create something positive out of the pandemic and teach our children ways of being part of a community and giving something back.

GIVE ME A BREAK... PLEASE

When a couple is each used to a busy professional live and have developed schedules that work for both of them as well as for their two young children, life is good. There is couple time together, professional time apart, Daddy time with the boys, and Mommy time with the boys, and family time. There is also going to school every day and a periodic child care as needed.

Suddenly, all four of them are sheltering in place and trying to construct a whole new life within the boundaries imposed by the COVID-19 pandemic. Mom was multitasking. The act of multitasking, resulted in not getting enough sleep, and was simply being exhausted. Dad conducted his practice at home using Facetime. Finding the quiet needed was a challenge in a small house. The biggest challenge was the noise created by two active boys.

After exploring various options, dad rearranged his schedule with clients so that each of them had some two or three hour blocks of time to simply give each other a break, even a long nap. Both parents found it personally helpful and helpful to their spouse to build in at least one long walk outside every day.

COVID-19 dramatically impacted every part of their lives. What made life better was their commitment as a couple to touch bases, keep communication flowing, and a willingness to make adjustments. Some of the things this couple took for granted have become new staples in strengthening their relationship.

FLOWERS ON THE STREET

Taking flowers to people in nursing homes has been something a particular lady has done for a long time. Every week, she goes to her grocery store and buys a large number of bouquets. Flowers are her way of bringing joy to patients as well as a way of "giving back" to her community.

One day, she bought bouquets of flowers. When she arrived at the nursing home, she was told that she could no longer deliver flowers because of the high number of nursing home patients in the United States who had contracted the COVID-19. No one, even family members, could visit patients. Sadly, she put the bouquets of flowers in the back of the car.

When she arrived home she decided to leave the large bucket of flowers on the sidewalk for people living in her neighborhood. Her neighbors were thrilled. Each week a new bucket of bouquets are left on the sidewalk. Neighbor after neighbor have thanked her and offered to pay for the flowers. Her response is touching, "Instead of giving me the money for your bouquet, please give the money to your favorite charity."

TEENAGERS MAKING A DIFFERENCE

Wash Hands, Wear Masks and Practice Social Distancing!

It should not be surprising how teenagers have taken an active role in helping others during the COVID-19 pandemic. Often, we adults do not give teens enough credit for their abilities, for their stepping up to help others, and for making a real difference in their communities. Perhaps in our busy lives, we have not taken time to listen, understand, and appreciate teens as they move towards becoming adult citizens of our nation. COVID-19 is teaching many of us a new respect for teenagers. When we acknowledge their efforts, we reinforce the kind of person they can become, including developing their better selves. Purposefully acknowledging our teens will also have the added benefit of helping us find our better selves.

LEAVE IT ON THE PORCH

In a neighborhood where a lot of senior citizens live, Kevin and Kavantre, two junior high school boys, came up with a great idea. Since seniors are high risk for COVID-19, the boys pick up small items for seniors from a nearby open grocery store or pharmacy. The senior citizen calls in an order and pays for it over the phone. One of the boys picks up the order, puts it in a backpack, bikes to the home, and leaves the bag on the front porch of the senior. Each boy tucks a note inside the bag. "We care about you. Stay safe. I am always available to help you. Have a good day." The boys have found a great way to help keep others safe by volunteering for an hour or two each day.

LET THERE BE MUSIC

In most communities, there are usually a number of good musicians that usually play in the school orchestra or jazz band. Friends in the orchestra and band came up with an idea to support people who are sheltered in place. Helen is quite a violinist and after dinner she goes around her neighborhood. She stops at each house, rings the doorbell and stands well away from the front door, wearing her mask. When the door is answered, Helen plays one or two pieces on her violin.

Robert plays the clarinet in the Jazz Band. The Jazz Band plays a lot of music from the 1930's and 1940's. Robert visits homes where seniors live, rings their doorbell, and plays songs from their generation. The smiles on the faces of seniors, as they remember when they were young, is heartwarming.

Anna comes from a large family and plays the guitar. She has learned to play children's songs for her younger brothers and sisters. Now she plays for about an hour in the afternoon, going from door to door playing and singing for young children who are at home.

CUTTING UP!

Rafael and John have fathers that work in the landscaping business. Each boy has access to a good lawn mower, edger, and burlap tarps. Their idea is to help senior citizens keep the front lawn of their homes looking nice. They work as a team, mow about ten lawns each Saturday and haul away the grass trimmings. Although the boys accept tips, mowing lawns for seniors is free.

FEED THE UNDERNOURISHED

Churches and community organizations often take on the challenge of providing meals for the poor and homeless. Some programs have existed for many years. Most often feeding programs are staffed by volunteers: retired teachers, nurses, social workers, as well as seniors from all walks of life. Senior citizens have more time to volunteer as well as a lifetime of amazing and helpful experiences. The system has worked well.

Enter COVID-19: One of the groups most at risk are senior citizens. Many seniors felt they could no longer come in contact with people who could, evenly remotely, have COVID-19. In addition, organizations feeding the homeless could no longer serve meals inside buildings.

What to do? Hungry people still need to be fed.

- Move feeding programs outside.
- Recruit volunteer teenagers to be incharge of the outside distribution of meals.
- Volunteer seniors prepare sack breakfasts and lunches in organizational kitchens the day before they are needed. Sack meals are stored in refrigerators.
- Completely sanitize outside tables before and after serving food.
- Mark sidewalks to help people maintain proper social distance.
- Volunteers always wear masks!

A young friend named Liam told me, "The need to keep feeding the poor and homeless is how my mother and I came involved in the STAR Program. As a junior high student, I like helping people who need food and they really like me."

STORY TIME

Suzie and Maria both have younger sisters and brothers that love to have stories read to them. Both use their voices very well, and add puppets and body movements when they read a book. Suzie and Maria have teamed up and recorded two stories that can be used just about the time most toddlers take a nap. Moms love the idea that someone else is entertaining their very young children, at least for a few minutes.

APPRECIATION SIGNS

A group of high school art students obtained poster boards to make "thank you" or "appreciation" signs. They contacted senior citizens and offered to make them a sign of their choice to put it in their front window. The seniors now have a way of giving thanks publically and the art students did a project that made a difference for the elderly in their community. Sometimes teens were actually able to place a "thank you" sign on the front lawn of someone who is an essential worker or medical staff member. This act is even more special because there is a direct connection.

Another group cut poster boards in half and delivered them to families with kids who are sheltering in place. Family members work together to create a "thank you" sign for their front window.

Contacting people they do not know is always a problem for teens. However, teens are good problem solvers. They start by asking mom or dad to help with people they know. Members of clubs, volunteer organizations, and churches often have rosters. It does not take many adults to help get things going.

DOG WALKING

Whether we are in a pandemic or not, dogs still need to be walked. Elderly people are at a high risk for contracting COVID-19 and, as a result, often elect not to leave their homes. A teen can be a welcome solution that is a genuine help to both dogs and owners. There are a few basic rules:

- Dogs must always be on a leash.
- Carry and know how to use "poop" bags.
- Stay clear of other dogs and people.
- Always wear a face mask.
- Thirty minutes is enough time for a good walk (Always check with the owner for the length of a walk.)

Walking a dog is not only good for the dog, it also helps teens burn off extra energy.

REPURPOSING POLITICAL SIGNS

Wash Hands, Wear Masks and Practice Social Distancing!

A group of junior high school friends spent a lot of time talking to each other on the phone and playing games during sheltering in place. One of the moms said to her daughter, "Why don't your friends do something positive instead of sitting around and yakking for hours on the phone?" The daughter did not take the comment well, but her friend on the other end did and came up with an idea. Their state had just had a primary election and there were campaign signs everywhere. The seven friends used "Zoom" to have a planning session.

The next morning each girl and boy in the group put on a face mask. Each one, individually, went around their neighborhood collecting as many old political signs as possible (In this state, election signs have to be removed within a week anyway). When the friends checked in, they had collected thirty-four signs. Step one of the plan was successful.

The next challenge was to find white paper to cover the signs. So many stores were closed, sheets of white paper were hard to come by. They told their mothers about their plan and enlisted moms to try and get butcher paper when they went to the grocery store. When the butchers learned of the plan, they gladly donated several feet of butcher paper. Step two of the plan was successful.

The group wanted the lettering to be brightly colored so messages could easily be seen. Group members shared ideas of where to find paint. Craft stores, they thought were sources for paint, were closed. Paint was available at big hardware stores, but who needs a gallon of paint for thirty-four signs? Finding paint, much less bright colors that would stand out, was not as easy as they thought. One of the dads overheard the discussion. He suggested they ask their dads if they had any partial cans of old paint in the garage. Almost everyone had old cans of paint and a variety of colors too. Step three of the plan was successful.

On their next "Zoom" call, they spend an hour discussing what should be painted on each sign. Then each member spent the day painting words on their signs and outlining each letter with a contrasting color. Creating the signs was more work than they imagined. Step four of the plan was successful.

That evening, the group discussed where to put the signs and how to coordinate time so that all the signs were placed within an hour or two. Step five of the plan was agreed upon.

Early the next morning, precisely at 7:00 AM, each member put on their protective face mask, gathered their signs, and headed down the streets. Most of the signs ended up on streets that are well traveled. By 7:45 AM, their mission was completed and they were back home. Step six: Mission accomplished!

What was on the signs? Each of the thirty-four signs thanked different groups for helping during the COVID-19 pandemic and for their long hours many were working.

THANK YOU
HEALTH
CARE
WORKERS

WE LOVE
NURSES
THANK YOU

THANK YOU
EMT'S
AND
FIREFIGHTERS

THANK YOU
TEACHERS
FOR
ONLINE LESSONS

THANK YOU
DAY CARE
PROVIDERS
FOR TAKING CARE OF
FIRST RESPONDER'S
CHILDREN

THANK YOU
HOSPITAL
CLEANING
STAFF

THANK YOU
SUPER MARKET CLERKS
FOR KEEPING FOOD
ON THE SHELVES

THANK YOU FRIENDS
FOR
SUPPORTING
FIRST
RESPONDERS

THANK YOU
FOR
SHELTERING IN
PLACE AND HELPING
US STAY SAFE

THANK YOU
ESSENTIAL WORKERS
YOU DO SO MUCH THAT WE
DON'T THINK ABOUT

The signs made essential workers and medical staff feel appreciated. There are a lot of people who are doing their jobs to their personal best everyday. Who are the people in your area that seldom get thanked? Although COVID-19 may be on the wane, many essential workers are emotionally exhausted after months of being on the front line of the pandemic. Thank you signs and cards are more important than ever.

Signs also have a huge impact on the community and get people thinking about others, not just themselves. We seldom realize the difference it makes when letting people know you appreciate them, especially now when so many people continue to give so much in combating COVID-19.

Life Lesson: It is amazing! When teens think about the needs of others, when they plan and work together, they accomplish absolutely amazing things. How often we sell teenagers short... their energy, enthusiasm, and creativity. When we adults support teens in their efforts to really make a difference, we are training the next generation of leaders, family, neighborhood, city, state, and national.

Younger Age Children

Wash Hands, Wear Masks and Practice Social Distancing!

Although many youngsters have started attending school again, many are on some kind of hybrid schedule. As students return to classrooms, what can we learn from the months and months they spent at home doing distance learning?

Shelter in place with younger children was a mixed bag. On one hand, they require more direct supervision, have a smaller attention span, and can get "whiney." Just ask any mom. On the other hand, they can be sweet, lovable, and helpful.

Most younger children have a lot of energy, need to change activities more often, and require more adult guidance: schoolwork, playtime, getting along with siblings, and helping around the house. Taking time to talk and plan with them each day can be a great help. So is having a written schedule/chart with spaces to fill in particular times and things to do each day. Think about how their day is scheduled at school and try to mirror it when they are schooling from home.

Appropriately involving younger children in simple decision-making can create buy-in, a sense of responsibility, and new habits that can last a lifetime. Below in one way it is done.

- Would you rather start school work at 8:30 AM or 9:00 AM?
- Would you rather set the table for breakfast or put the dishes in the sink after we eat?
- When you have a break, what game would you like to play in the backyard?

At school, television is not on during the day. The same should be true at home.

We know that some students were able to maintain their level of learning, the academic progress of other students suffered. So much of student success, especially young students, depended on the availability of technology at home, the parent's knowledge of technology, and the availability of a parent to guide their child's learning during the day.

MATHEMATICS AT HOME

Whether it is distance learning or packets sent home, most schools provide specific assignments during these pandemic times when schools are totally or partially closed.

When a child is learning a mathematical skill or operation, it is helpful for an adult or teenager to be there. Sometimes, the person answers questions or explains things using different words. If the child is having difficulty completing an assignment, try to come up with a different way or example to explain what the child needs to know. Older children can also help younger siblings as well as make flash cards.

Being at home a lot provides opportunities for a child to learn about the practical side of mathematics, how mathematics is used every day in real life situations. Some activities around the house might include youngsters walking around the house and making a list of everything that is a rectangle. Another time they can make a list of circles, squares, arcs, triangle. Similar activities can be done outside. These are good ways to help youngsters see how math and geometry are a part of everyday life. Another activity is to have a discussion on what kind of jobs require a knowledge of mathematics.

- How do doctors use math?

- How do carpenters use math?
- How do computers use math?
- How do cooks and bakers use math?

MATHEMATICS AND COOKING

(Source: Meril Smith)

Many families have a children's cookbook tucked away somewhere. I still do not know where ours is, but I have seen it in the past when grandchildren visit and are excited to make something with grandma. Learning to read and follow recipes is an easy way to help children really understand that math is used all around them. Grandmas are often perfect kitchen math teachers.

Parents are often just plain too busy to spend the time helping their children learn how to cook. It's a great time to find out why they like certain foods and what kinds of foods you ate as a child. Sheltering at home provided unique opportunities for child-adult time in the kitchen.

- Do some simple cooking together. Just learning the basics of preparing a meal is a good thing.
- Mathematics can be easily included if you use a cookbook. A child's cookbook is a great way to start learning—a cup of (8oz), two tablespoons of, how many teaspoons are in a tablespoon, how many cookies will this recipe make? How many cookies are in a dozen? When cooking pasta, what size pot is recommended and how much water should be in the pot? You'll come up with even better ideas.
- A ruler or tape measure is another way to help youngsters understand how we use mathematics in our everyday lives. Some examples might include measuring the depth of a kitchen shelf. Will dinner plates fit on the shelf? Is there any extra space? Measure the height of the kitchen table and the counter top. If they are different, why? What is the width of various doors? Why might some be different? Older kids can use a tape measure to find out the area of a drawer, etc. Why are some drawers larger or smaller or are deeper?
- How many gallons will the kitchen sink hold, the bathroom sink?
- Choosing one different practical math activity each day, can create a real understanding of how math is used in our everyday life.

JOURNALING

Journaling is something that youngsters of all ages can do. Journaling can also become a lifetime habit. The COVID-19 is a stressful time for everyone, including children. It's hard for all of us to see unexpected changes happening in our lives, especially changes that affect everyone around us. Journaling is a good way to put stress into words as well as give children a way to express how they are feeling each day about what is happening around them.

- Younger children draw a picture of something that happened each day. Date each picture and have your child tell you about their picture. You might want to write one or two sentences at the bottom. Clip them together to share later.
- Grades 1-3 can draw a picture on the top half of a page and write from one sentence to a three or four sentence paragraph. The following is one way of helping a youngster develop a paragraph.

Start with the topic sentence

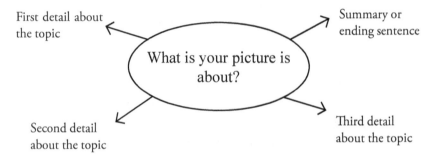

First detail about the topic

Summary or ending sentence

What is your picture is about?

Second detail about the topic

Third detail about the topic

Drawing and writing are good ways for youngsters to express how they are feeling, especially in a time of stress. Parents can also ask simple questions to help kids open up and how they may be feeling.

- Can you tell me more about this idea or this part of a picture?
- This is an interesting word. Can you tell me more about what you were thinking?

Date and clip each story together so they can be shared later, especially when they become adults!

HAND PRINT NOTE CARDS

Another great project for younger children is making "hand print" cards using tempera or other water soluble paint. Brush a little bit of paint on their hands at a time. Kids tend to want to stick their whole hand in paint. Try handprints out on a piece of newspaper or other scrap paper until they get an idea and find a "look" they like. The same idea can be used to make wrapping paper.

WRITING APPRECIATION NOTES

Many younger children know the name of someone who is really important to them. Children can certainly write a paragraph to tell these people how much they are appreciated. After the first letter or two, children can begin to write for exhausted medical people who have worked so hard for so many months. As COVID-19 become less of a threat to masses of people, an appreciation note is often just what many exhausted first responders need.

GET WELL CARDS

Being in the hospital or a nursing home is bad enough, but when family members cannot visit, it is ten times worse. Young children can bring a smile to a face of a senior citizen when they receive a get well or greeting card from an elementary school age youngster.

Four to six cards can be made by young children every few days as an on-going project. The art work is uplifting and a few words about getting well are enough. Second through fifth graders can write a message or make up a poem.

A parent can make a phone call to find out how to address a large envelope, containing all the cards, so they will get to the patients. Get well cards are not just for COVID-19 patients, but also for people in the hospital for lots of different reasons.

Such a simple thing as making and sending a card really helps a child develop a sense of empathy and caring that will last a lifetime. We all need to remember that each of us will get old and what it will be like to know some young people really care about us.

RAINBOWS ON THE WINDOW

Rainbows are a traditional symbol of better times ahead. Young children can create a rainbow picture and put it on the front window. They can also make a rainbow get well card. A bit of science can also be taught. A rainbow is simply light broken into component colors through raindrops or a prism. The colors in a rainbow are always arranged the same: red, orange, yellow, green, blue, indigo, and violet.

SOCK PUPPETS

Most of us have a one-of-a-kind socks in the back of a drawer or box. Making sock puppets, using a variety of different socks, is a great activity. Socks can be decorated with felt tip pens, buttons, etc. Since there are various sizes of socks in most homes, a sock family can be created. Everyone in the family can have a different sock puppet and make up stories or plays together. Sock puppets are an easy way to help children explore their feelings, their fears, and the uncertainties they face growing up.

COLLAGES

Old magazines are a great source for creating a collage. Topics could be animals, trees, sports, toys, of anything that interests a child. Going through magazines and cutting out pictures can take a good amount of time and putting a collage together with paste or glue can take even more time.

Another kind of collage can be made from plastic milk jugs. With a bit of help and guidance, a plastic milk jug can be cut out into the shapes of various animals: pigs, elephant, bear, dog, etc. Cutting out shapes of colored paper and gluing them on the milk carton gives texture as does adding things such as pipe cleaners. This project does require some discussion about animals, shapes, and how to cut out legs and other parts. Working with a plastic milk jug is good project for a dad and child working together as a team.

A BACKYARD PAR COURSE

(Source: Gina Foster)

Liam's and Izzy's mom made a par course in the backyard. The boys run, jump, balance, serpentine, as well as do sit-ups, jumping jacks, waist twists, bouncing a ball the same number of times as their age, and making one last shot into a bucket. (The ball in the bucket leaves the activity ready for the next person.) After an hour of school work, spending fifteen minutes on the obstacle course can be a welcome break.

HOPSCOTCH

Using sidewalk chalk, it's easy to make an old fashion hopscotch game on a walkway or patio. Hopscotch is a great game to develop balance and bending skills. It's also a game that siblings can play together.

JUMP ROPE

Jump rope is another traditional game. If you have several children, the older ones can teach younger ones how to jump rope. Jump rope is a great way to develop coordination skills.

MAKING AN INSIDE "HIDEOUT"

As kids, many of us made a tent inside the house. Sheltering in place is an opportunity to expand on the idea. Use an old sheet (the bigger, the better) as the covering for the tent. Work in a place where crayons, felt tip markers, colored pens, etc. will not transfer onto carpets or floors such as a patio or garage floor. If you have several children, divide the tent into sections for each youngster. It may take a couple days to do the actual art work. Set up the hideout using chairs, tables, or other items around the house as a frame.

PLAY-DOH

If you do not already have play-doh at home, there are several easy recipes online. Adding a drop or two of food coloring and letting kids kneed and store play-doh in small containers with lids. It's a great way to have an extra activity tucked away just in case you need one.

RIVER ROCK PAPER WEIGHTS

(Source: Meril Smith)

Smooth river rock about the size of a youngster's hand makes a great paper weight gift. Painting an animal, flower, or name is a good activity requiring the use of small hand muscles. There may be a few leftover pieces of kitchen tile in the garage that make a great base when the rock is glued on.

THE ALPHABET GAME

The alphabet is a simple game that helps young children think of words by categories. It can also be played by a team of a younger and older child, or by two teams of a parent and child.

ALPHABET GAME

	Name	Food	Toy
A	Angie	apple	airplane
B	Ben	berries	boat
C	Carrie	carrots	car
D	Danna	dates	
E	Emmy	endive lettuce	
F	Francisco	french fries	
G	Geri		
H	Henry		
I	Izzy		
J	Juanita		
K	Kerry		
L	Lucy		
M			
N			
O			
P			
etc.			

VEGETABLE GARDEN

A vegetable garden can be a small plot of land next to a fence, several larger flower pots, or even a larger area in the backyard. All these areas can be used to grow vegetables. A vegetable garden can be a family affair or each youngster is assigned a small area or a couple of containers.

Find out which vegetables are easy to grow in your area and the time of year they should be planted. Preparing the soil, hoeing weeds, turning the soil using a shovel, making sure the soil is loose and not packed, making furrows, are just some ways of getting the soil ready for seeds. Be sure the youngsters read and follow the directions on the package. Finally, create a log from planting to harvest.

Sample Garden Log

Activity	Date	Seeds	Plants	Height
Prepare soil	5-25			
Plant	5-27	Bean Seeds	Tomato Plants	3" high
Schedule	5-28			
Watering	6-13	First sprout	Tomato Plants	5" high
Every other	6-20	1 " high		
Day	6-30	3 " high	Tomato flowers	7" high
	7-15	Flowers	Sm. Tomatoes	8" high
	-----		First Harvest	

When youngsters actually grow vegetables and prepare them, they are far more likely to eat them, even ones they may not like.

WEED PULLING CONTEST

Having three children at home 24/7 is a "stretch" for any parent. Working together, children and parents, can create new activities and ones that can be a challenge.

About six week into shelter in place, one mom, with three young kids, was ready to start "pulling out her hair." Her frustration gave her an idea, pulling out the weeds. The ground was fairly soft due to some recent rains and the time was right. Here are the rules of the contest.

- The contest lasts for only 15 minutes.
- Each child gets a box for the weeds they pick.
- To count, the weed must have the root attached.
- Size does not count, only weeds with roots count.
- Each child counts the weeds in front of mom and the siblings.
- The winner gets to choose the family game for the evening. (I am sure you can come up with other prizes.)

The game worked. Mom got away with the kids playing the game about three times. Then the kids caught on. All good ideas come to an end. I am sure you can come up with equally good ideas...just do not pull out your hair.

DADDY LED ACTIVITIES

WOODEN TOYS

(Source: Meril Smith)

Some dads have shops in the garage, others may have a handsaw, a drill and sandpaper. Most dads have a few pieces of scrap wood. With a little imagination, youngsters can make their own wooden toys. Toy kits can also be ordered online.

The three items above are samples of hundreds of toy kits that are available online if dad is not into woodworking. When I made toys with my grandchildren, we always signed the bottom with our names and the date. Twenty years later, the grandkids still have their toys that show a lot of wear and tear.

BOYS NIGHT OUT

Night out is not a night out with the "Boys." Rather it is a night out with your kids, boys and girls. If you have a tent, great! If not, a couple of large blankets over two ropes work just as well. If you have a tarp of some kind it helps keep the night moisture away. Sleeping bags or heavy blankets both work for sleeping, if you have air mattresses, even better. Flashlights, story books and games are a must!

A BBQ with hamburgers, hot dogs, and grilled vegetables make a great dinner for the campers. An outside dinner also gives dad the opportunity to cook with the kids.

When it is time to go inside the tent, a whole new adventure can begin. Using your fingers, you and the kids can make shapes like animals. With the flashlight on one side of your hand, the shape will have a shadow on the side of the tent. Another activity is to create a round robin story. Dad starts a story, followed by each youngster adding to the story. Funny stories and ghost stories both work well. In addition, games like checkers, Uno, Sorry or another favorite game can be played before bedtime.

Dad "camping" with the kids will create a long lasting memory. Years later your children will say, "Dad, remember when we..."

LOSING AT GAMES

Children love playing games, especially seeing "Dad lose." Whether it is a game like checkers, Uno, or Fish, kids love seeing a parent make a "big" mistake or losing. My grandchildren love seeing me lose. It is even more fun when the act of losing is very animated. "Oh no! Did I really do that?"

SMALL BUSINESSES
Wash Hands, Wear Masks and Practice Social Distancing!

Most people have taken the COVID-19 pandemic and restrictions from their governors very seriously. Sheltering in place, wearing face masks, and practicing good hygiene were not easy for small businesses. It is also a good feeling that the United States may have turned the corner on COVID-19 and that many businesses are able to reopen even with some restrictions. However, the COVID-19 continues to rage in other parts of the world and will certainly not completely disappear from the United States for a long time. We Americans should not forget what we have learned, how we can rise to any crisis, how we actually can adapt to new situations and actually come out better.

When COVID-19 hit hard and was totally out of control, almost all businesses closed and waited until being told when they could begin the reopening process. To say this was hard on small business, and small business owners are a gross understatement.

Some small businesses used telephones, computers, and home delivery to serve customer needs. Other small businesses, often restaurants, developed take out systems. Most owners of small businesses struggled to survive and many did not survive.

There is a bright side to all this. There are wonderful stories of how small businesses came up with creative ways to help bridge the COVID-19 closure gap. Many of these stories are worth retelling.

FEEDING HOSPITAL WORKERS

This story took place at the very beginning of the pandemic.

In large industrial areas, there are often restaurants that thrive by providing breakfast and lunch to hungry workers. Some restaurants actually serve hundreds of meals a day and go through tremendous amounts of food. Restaurants usually order food from wholesalers twice a week.

One such restaurant had just received a large food order. The next day it was ordered to close as a result of the rapidly spreading COVID-19 pandemic. The storage shelves were full. The cold storage unit was full. The freezer was full.

As the restaurant owner watched the havoc that the COVID-19 was having on his community and the health care system, he was in awe of something that had been incomprehensible just a few days before. He saw pictures of health workers on television that looked completely exhausted from helping very ill patients and how they were reacting as the number of patients dying increased at an alarming rate. A friend was hospitalized with the COVID-19. He knew he had to do something to help!

Realizing that his restaurant was fully stocked with food, the restaurant owner contacted the hospital. The hospital cafeteria was shut down. Quickly, a plan was created to serve free hot meals to over a hundred hospital workers two shifts a day—doctors, nurses, paraprofessionals, laundry staff, and cleaning staff—until the restaurant ran out of food. To help, four of the restaurant employees were brought back to work.

However, the food did not run out. Once the local television station heard about the free meals for health care workers, they aired the story. For months, additional food stocks and money have been provided by individual donations and wholesale businesses. In times of great crisis, this story is an example of one person seeing a need, stepping up to the plate, and making something positive happen.

A STORY TOO GOOD NOT TO SHARE

The Inn at Little Washington, a famous three-star Michelin restaurant in Virginia, approached reopening for inside dining in a wonderfully unique way. Their idea may spark hundreds of other ideas around the nation that may help restaurants financially survive during the lengthy COVID-19 pandemic.

The Inn at Little Washington created a way to help diners abide by the social distancing guidelines while enjoying dinner in a unique environment. Diners are seated next to completely set tables with beautifully 1940's style dressed mannequins.

Chef Patrick O'Connell said, "This would allow plenty of space between guests and elicit a few smiles and provide some fun photo ops."

THE 1882 GRILLE

A small city in Oregon prides itself on its beautiful tree-lined vintage main street and its variety of local shops and restaurants. Customers and shop owners actually know each other. Residents take pride in the fact that no chain stores are allowed in the downtown area and people go out of their way to support "Third Street Merchants."

The 1882 Grill is a popular family restaurant on the third floor of a quaint old building. Its location requires most customers to use an elevator. The food is excellent, the prices are reasonable, and the view of the surrounding downtown area is spectacular. Like most restaurants, COVID-19 forced the 1882 Grille to shut their doors. Only a few restaurants were easily adapted to a "Take-Out Only" situation. However, doing "take-out" from the third floor? You have got to be kidding!

As government and health officials thought they were getting a handle on COVID-19, discussions were started about how to do a careful, one small step at a time, reopening of some businesses. Some shops were open to one customer at a time. A customer waited outside the door wearing a mask and gloves. Social distancing lines were marked on the sidewalks. When one person left the store, another person is allowed to go in. The process worked well. However, this approach did not seem viable for a restaurant on a second floor.

Finally, a plan was formed to reopen 1882 Grille, DRIVE BY! The staff selected a few entrees from the menu that could be adapted to take-out. The 1882 Grille used Facebook to contact as many customers as possible. Customers could call in orders for two or three of the entrees and pick them up between 4 P.M. and 6:30 P.M. for their reopening. Dinners were free on opening day. Community response to the reopening of the 1882 Grille was beyond anyone's imagination.

On Saturday, six long folding tables were put at the curb of the street. 7,500 meals were nicely packaged and put into labeled paper bags. Bags were brought down the elevator to the tables. About 2000 bags sat on curbside tables waiting to be picked up.

Restaurant staff greeted people in each car, checked their order. Of course, everyone was wearing brightly colored masks. Firefighters and restaurant staff put dinner bags into each car through the passenger side window. There were so many cars that members of the police department directed traffic. They marked one street for driving to the restaurant, closed a side street for the actual pickup, and then directed traffic when and how to turn onto the main street of town to leave.

When a whole town decides a community business is really worth saving, amazing things actually happen! The 1882 Grille was saved, continues to take phone orders and is now open for diners.

Incahoots

Incahoots is a popular little shop where you can sit and have tea and shortbread, buy a beautiful bouquet of flowers, or shop for gifts and garden plants. COVID-19 shut them down immediately, No customers, no tea, no income! Incahoots is noted for creativity. On the parking lot side of the building there is a storeroom with a fairly large window. Idea! Take everything out of the storeroom. Install a counter outside the window. Create a drive-through. Incahoots is open for drive-through business including tea and shortbread and has reopened for limited inside business.

When we learn to think "out of the box," new solutions often happen, even in a storeroom. Way to go!

HOPSCOTCH TOY STORE

Kids, parents, and grandparents love Hopscotch Toy Store. It's a locally-owned shop that really focuses on kids. The mix of toys and games caters to a wide variety of ages and interests. When a youngster is going to have a birthday party, Hopscotch will help the child put together a basket of party prizes.

Naturally, as a result of the COVID-19 pandemic, Hopscotch was closed. The business has survived because of phone orders and delivery service. With the partial lifting of restrictions, Hopscotch has reopened as a "take-out." Hopscotch is located in an old building on the main downtown street. It has two doors. One door remains locked. The other door is open and the owner has put a large desk across it. While wearing masks, customers can walk up to the desk. While social distancing, toys can be purchased or picked up.

(source: Meril Smith)

If it is the day of a birthday, the child and his mother are allowed to come inside the story to pick out toys. Of course masks and social distancing are required.

Izzy was celebrating his sixth birthday. His excited comment to his mom was, "Do we really get to go inside Hopscotch because it's my birthday? Yippee!"

Hopscotch is now open for people being inside. This story is another example of a small business dealing with the COVID-19 Pandemic for over a year and SURVIVING.

DINING OUT:
A DIFFERENT APPROACH

Despite financial help from the Federal Government, there remains concern that hundreds of thousands of restaurants across the country will not survive the economic ravages of the COVID-19 pandemic. Even when allowed to reopen, there may be restrictions on the number of customers allowed. One estimate is that many restaurants may end up with a new capacity of only twenty-five percent, which may not be enough to survive.

(source: Gino Foster)

McMinnville, Oregon has a wonderful downtown street that is reserved for only small, independent businesses. Chain stores are all located away from downtown. Except for a few restaurants who offered take-out, COVID-19 shut down the entire street.

As restrictions begin to relax just a little, some business people started discussions about actually closing the main street each weekend and making it into an outdoor "eating mall". The concept allowed the many restaurants to expand their capacity out onto the actual street and dramatically increase business while maintaining "social distance" between tables.

In other small cities, where street malls are permanent, different kinds of planters and other unique barriers have been added to create a wonderful outdoor dining experience. Because of many people dining on the main street, other types of business greatly benefited from the relaxed atmosphere and additional foot traffic.

When this story was written during the first few months of the pandemic, I wondered if this idea would actually be implemented.

The bottom line is that the McMinnville Downtown Association actually did implement the idea. The idea of changing Third Street into a restaurant and shop mall worked so well they are closing Third Street again on weekends throughout the spring and summer.

TRY THINKING INSIDE THE BOX

Wash Hands, Wear Masks and Practice Social Distancing!

Most of us know the expression, "Try thinking out of the box." Certainly, the COVID-19 pandemic has challenged each of us to "think-out-of-the-box" as we have drastically changed the way we go about our daily lives. However, at this time, we also need to learn to think "inside-the-box," both figuratively and literally.

The odds are poor, at best, that we have seen the last of the COVID-19. There is a good possibility we will see new mutations, variants, and mini spikes for months and months to come or in areas where people have let down their guard and have assumed the worst is over.

Lest we forget: the Antonine Plague around the Mediterranean Sea lasted from 165 to 180 AD. One third of the population living around the Mediterranean Sea died. It is hard to believe that today's officials and scientists did not come out stronger about the history of plagues and pandemics. Good information might have made a difference. America certainly had more than enough historical and scientific information about the 1918 pandemic. There is an old saying which I truly believe: "If we do not learn from history, we are condemned to repeat it."

Each family is different—different sizes, different ages, and different stages. For sure, families are not a "one size fits all" proposition. This section contains ideas for coping with additional wave and mutations, as well as information that may help families keep life during pandemic in some kind of perspective.

FIRST, OUTSIDE THE BOX

The COVID-19 pandemic has forced all of us to slow down. It has forced us to reexamine our priorities and how we interact with others. The pandemic affects the entire world. We are all under attack. COVID-19 is an equal opportunity virus. It doesn't separate people by class, race, religion, politics, or where we live. The entire world population faces this pandemic together.

As parents, we set the example for our children. We can demonstrate doing our part every day: even if it means sheltering in place again, wearing face masks again, and washing our hands many times a day for 20 seconds again. More importantly, our children see and learn how we treat other people. They see if we make frequent phone calls to family members, just to say hello and check to see if they need help. Our children see if we reach out and help elderly or people who may be at high risk for the virus. Our children also see if we reach out and help our community get through the pandemic.

Indeed, we are the role models for our children. What we do now, what we say, and how we behave towards others, especially during the COVID-19 pandemic, is what our children internalize and how they will approach their adult life. The COVID-19 pandemic is a defining moment for our children's generation.

This is the time for parents to think out of box for the future of their children. This time is a unique opportunity to shape the values of the next generation of adults.

SECOND, INSIDE THE BOX

COVID-19 spikes are likely to be with us for a long time to come. Spikes may be a result of new variants, geographic areas where people choose not to be vaccinated, or individuals becoming complacent and participating in high-risk behaviors.

More importantly is to realize that few of us plan ahead for major unexpected events or disasters: hurricanes, tornadoes, floods, earthquakes, epidemics, or pandemics.

How many of us actually have an emergency kit with three to five days of water and food? In my home, our emergency food supply is neatly tucked away in a box with wheels. It is in a corner of the garage and we seldom even notice it. However, it is there if we ever need it. Where do you keep your emergency food kit?

However, we were certainly not prepared for anything like sheltering in place for weeks. As a result of COVID-19, we now know that the need for a different kind of emergency kit has emerged, pandemic first aid kits.

"What you say? What is a Pandemic First Aid Kit?" These kits are about what each member of the family might need in an emergency, over and above food and water, to keep them safe and busy for an extended period of time. Looking back at the months and months of COVID-19, what do you wish you had had in the house that you did not have? I have heard from moms that they wish they had put together a little survival kit for each of their children. So, let's create "Pandemic First Aid Kits."

First off is a cardboard box – a banker's box or larger. Hopefully, it has a lid.

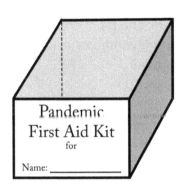

You may have several items on each list around the house – don't rebuy. It's also not necessary to put everything in the kit. Just add them onto the list so you will not forget when you need to remember.

MOM FIRST!

MOM'S
PANDEMIC
KIT

As a mom, what would you include in your own Pandemic First Aid Kit? What will make it easier to survive with your kids and husband? Place a few special items in your Pandemic First Aid Kit.

- Two of the largest bars of dark chocolate available
- Two extra-large rolls of toilet paper – just for you!
- Two large bottles of hand sanitizer
- A whole box of face masks – just in case

- Pens, paper, envelopes and stamps
- Favorite soap, hand cream, and night cream
- A couple of good books or favorite magazines
- A favorite card or board game
- Two or more favorite music CDs.
- Items you need for your favorite hobby
- A personal calendar and personal phone book
- Stay connected with friends
- Favorite ways to relax

Mom, the most important thing is: Plan now on how to take care of you!

PANDEMIC FIRST AID
KITS FOR KIDS

Ideas for ages 3 – 7 (Some beginning ideas, you will have better ones.)

- A couple of card games they like...Uno, Go Fish, etc.
- A couple of board games...Candy Land, etc.
- A couple of favorite books
- A new card game, board game, book at their level.
- Math or Sight Word Flash Cards
- Math, Language Arts, Science Workbooks (Teacher Supply Store)
- Easy Word Search Book
- Large roll of white shelf paper
- Washable colored markers or felt tip pens
- Box of large, thick crayons
- Box of sidewalk chalk
- Tablet of alphabet letters... to practice printing
- Large size Index Card... for making greeting cards
- Big roll of scotch tape
- Six primary grade sized pencils

- Scissors
- Paste or glue stick
- Old magazines with a lot of pictures to cut out
- Do a computer search for age appropriate games

Ideas for ages 8 – 12 (Beginning ideas, you and the kids will have better ones.)

- A couple of card games they like
- A couple of board games
- A couple of favorite books
- A new card game, board game, or book at their level.
- Math, Language Arts, Science Workbooks (Teacher Supply Store)
- Word Search Book
- Easy Cross Word Puzzle Book
- Large roll of white shelf paper
- Washable colored markers or felt tip pens
- Large box of crayons
- Easy learning to draw book
- Large size drawing pad
- Box of sidewalk chalk
- Large size Index Card...for making greeting cards
- Big roll of scotch tape
- Scissors

Ideas for ages 12+ (Some ideas, your kids will have better ones.)

- A couple of card games they like
- A couple of board games
- A new card game, board game, book
- Math, Language Arts, Science Workbooks (Teacher Supply Store)
- Word Search Book
- Word Puzzle Book
- Large roll of white shelf paper

- Washable colored markers or felt tip pens
- Large size Index Card for making greeting cards
- Scissors and a big roll of scotch tape

You may have some of the things on each list already around the house. It's not necessary to put everything in the kit. Just add them to the list inside the lid so you will not forget when you need to remember.

It would be great if you did not ever need to use your Pandemic First Aid Kits. You can always save them for cold days during winter break or on hot days during summer vacation.

PANDEMIC KIT FOR DAD

A Pandemic First Aid Kit for myself? My first thought was a few bottles of... My next thought was a complete set of hair clippers. (I almost had to get a dog license during the first shelter in place.) How about a game that both my wife and I enjoy playing? I am also thinking about adding a game I would like to teach to my children and grandchildren. Let's include two good books that I have put off reading for the past couple of years. Finally, I'll put in a copy of my favorite saying.

NO ONE CAN DO EVERYTHING.
EVERYONE CAN DO SOMETHING.
WORKING TOGETHER,
WE ACCOMPLISH ABSOLUTELY
AMAZING THINGS!

> **Life Lesson:** Pandemic First Aid Kits for each member of the family are not just a good idea for now, they are good to have around... maybe stored in the garage. It's nice to know you have something at your fingertips just in case of some other crisis.

DON'T FORGET!

Don't forget that COVID-19, in some form, will be around
for several years...diminished but not gone.

Don't forget that you matter,
your children matter, your family matters.

Don't forget there a good reason to be cautious
until COVID-19 is fully under control.

Why not continue to "Wash Hands,
Wear Masks, and Practice Social Distancing?"

Why take the chance when it comes to COVID-19?
If you do not like wearing a mask, you are
really going to hate a ventilator.

Most of all, "Be Safe!"

Life Lesson: In a time like this, people can find their better selves, model compassion, empathy and help others. We are all in this pandemic together. May we learn to see how much we are alike and not so different. May we see the importance of being a part of the greater community and caring for others as well as ourselves.

The truth is that we are our children's first and most important teachers. May we teach and model our better selves today and everyday. What we, parents, say and do will last our children a lifetime. This is the way we and our children truly become our better selves.

A FINAL STORY:
WE ALSO NEED TO BE
ABLE TO LAUGH

The first year and a half of the COVID-19 pandemic, it was often hard to find much humor. Dr. Anthony Cedoline, psychologist and author, sent me the following list which helped me put things into perspective as well as help me smile and laugh more. Maybe you will enjoy this story too.

MY 56th DAY OF ISOLATION
(Anthony Cedoline, PhD)

As a psychologist, I spent a career working with adults and children. However, when the COVID-19 became a pandemic and shelter-in-place continued for many weeks, I began seeing new kinds of stress and challenges. Here are some of the email comments I have received:

Day 7: Okay, so schools are closed. Do we drop the kids off at the teacher's house?

Day 9: I am homeschooling. The first day I tried to get my kid transferred out of my class.

Day 13: It may take a village to raise a child, but I swear it's going to take a vineyard to homeschool mine.

Day 21: I hope they give us two weeks notice before sending us back out into the real world. I think we'll all need time to become ourselves again. And by "ourselves" I mean lose 10 pounds, cut my hair, and get used to not drinking at 9:00 AM.

Day 30: I am concerned about my new monthly budget: Gas $0, Entertainment $0, Clothes $0, Groceries $2,799.

Day 32: Breaking News: Wearing a mask inside your home is now highly recommended. Not so much to stop COVID-19, but to stop eating.

Day 33: We, low-maintenance women, are having our moment right now. We don't have nails to fill and paint, roots to dye, eyebrows to re-ink, and are thrilled not to have to get dressed every day. I have been training for this moment my entire life!

Day 35: When Shelter in Place is over, let's not tell some people.

Day 36: I stepped on my scale this morning. It said: "Please practice social distancing. Only one person at a time on the scale."

Day 39: Appropriate analogy: The curve of new COVID-19 cases is flattening, so we can lift restrictions now. Or could it be said, "The parachute has slowed our rate of descent, so we can take it off now."

Day 40: They say that things may open up soon. I'm staying in until after the fourth wave of COVID-19 to see what happens to you.

Day 44: People keep asking: "Is coronavirus REALLY that serious?" Listen carefully, the churches and casinos are closed. When heaven and hell agree on the same thing, it's a good bet that COVID-19 is serious.

Day 46: Never in a million years could I have imagined I would go up to a bank teller wearing a mask and ask for money.

Day 52: Not to brag, but I haven't been late to anything in almost eight weeks.

Day 54: I was in a long line at 7:45 AM waiting for the grocery store to open at 8:00 AM. Tuesday and Thursday mornings are for "seniors only." A young man came from the parking lot and tried to cut in at the front of the line, but an old lady beat him back into the parking lot with her cane.

He returned and tried to cut in again, but an old man punched him in the gut, then kicked him to the ground and rolled him away. Bruised and battered, he approached the line for the third time. He said, "If you people don't let me unlock the door, none of you will ever get to shop for your groceries."

Day 56: For the next wave of the COVID-19 pandemic, do we have to stay with the same family or will they relocate me?

These comments resulted in a variety of emotions. They also have given me plenty of things to ponder. Whether we like it or not, I know that all of us are truly in this pandemic together. The best way to get to a new normal is to practice the precautions we have been told to practice time and time again. We will COVID-19! What is required is that all of us think of others and that we do our part to create the best possible outcome.

> *Life Lesson: Sometimes we take situations so seriously, that we forget to see humor in even the most bizarre situations. Stepping back from the absurdity of the pandemic helps us survive well emotionally. After all, laughter is the best medicine.*

About the Author

Meril Smith grew up at the end of World War II with children of migrant farm workers, with children who were born in Japanese internment camps, and with children of day laborers and blue-collar workers. Poverty, recessions, and helping each other were all basic parts of surviving in his young world.

Living through the times of the Berlin Wall, epidemics of polio, measles, chicken pox and mumps as well as economic recessions, the assassination of President Kennedy, the space race, the development of Silicon Valley, the Vietnam War, and the tragedy of 9/11 have all fueled Meril's passion for understanding people and events.

HAVE WE FOUND OUR BETTER SELVES? is a follow-up to a "survival" book that was written during the first three months of the COVID-19 pandemic. With over a year and a half of living with the COVID-19 and with the probability of COVID-19 finally being somewhat under control, this book helps all of us step back and take a look at how we have thought and acted during the months and months of the pandemic. Ask yourself, "Am I just trying to forget the whole thing as if it were just a nightmare? Or, are there things I have learned about myself and my family that are important now and in the future? What new skills have I learned? Are there ways I think about and treat other people that have changed? Do I like myself better? Am I finding my better self? Am I also helping my family find their better selves?

Meril's unique experience as a spacecraft controller during the moon landings, a love of history, an enjoyment of research, a good sense of humor, a wide variety of life experiences, and a forty-year career in all levels of education provides an excellent backdrop for Meril's informative and easy to read books.

CPSIA information can be obtained
at www.ICGtesting.com
Printed in the USA
BVHW071159310821
615693BV00010B/505